ANIMAL EATING HABITS

THINGS ANIMALS DO

Kyle Carter

The Rourke Book Co., Inc.
Vero Beach, Florida 32964

Edited by Sandra A. Robinson and Pamela J.P. Schroeder

PHOTO CREDITS
All photos © Kyle Carter except page 21 © Tom and Pat Leeson

Library of Congress Cataloging-in-Publication Data

Carter, Kyle, 1949-
   Animal eating habits / by Kyle Carter.
   p. cm. — (Things animals do)
   Includes index.
   ISBN 1-55916-111-6
   1. Animals—Food—Juvenile literature. [1. Animals—Food.]
I. Title. II. Series: Carter, Kyle, 1949-   Things animals do.
QL756.5.C38 1995
591.53—dc20   94-46846
                                          CIP
                                          AC

Printed in the USA

TABLE OF CONTENTS

EATING

You may have better table manners than wolves, but you still have something in common with them. You have been just as "hungry as a wolf."

Wolves and other animals eat for the same basic reason you do—to survive. Without food and the energy it provides, living things die.

Animals sometimes find plenty of food, but often they do not. Many animals have a full-time job finding or catching food. Some dive or dig for food. Some run, walk, wait or climb for it.

Wolves fill themselves
quickly on a deer dinner

WHO EATS WHAT AND WHOM

Animal appetites are as different as the animals themselves. Exactly what an animal eats depends upon many things. Most animals are somewhat fussy about what they eat.

Usually, animals live on a diet of plants *or* animals. For example, deer and rabbits are plant-eaters, or **herbivores.** Lions are meat-eaters, known as **carnivores.** They are predators—they hunt other animals for food.

A few animals, such as raccoons and most bears, are **omnivores.** They eat both plants and animals.

Raccoons have a great sense of touch, so they feel.for food of all kinds with their paws

LIMITS ON APPETITES

Animals aren't fussy by choice. They are born with built-in tastes. In addition, what they eat is limited by what they *can* eat.

The lion has no appetite for a meal of plants, and it doesn't have the kind of teeth needed to cut and chew them. A hungry lion won't fill up on leaves or grass if it can't find **prey.**

An osprey, or fish hawk, lives entirely on a diet of fish

TEETH AND BEAKS

Teeth and beaks tell much about an animal's diet. Almost all reptiles, such as the crocodile, have mouths full of sharp teeth, or they have sharp-edged jaws. Nearly all reptiles eat meat.

Meat-eating mammals—lions, tigers, seals, badgers and others—have sharp teeth, too. Plant-eating mammals usually have broad, flattened teeth.

Beaks are also matched to appetites. A heron's long bill is perfect for spearing fish. A parrot's wide, heavy bill is ideal for crushing fruits.

The crocodile's mouth—the mouth of a predator—is made for holding and tearing flesh

*Without teeth to chew, birds of prey—like this young bald eagle—
swallow large chunks of meat*

A wood stork coughs up "fish stew" for its chicks

FINDING FOOD

Most baby animals don't have to hunt for food. A parent brings food to them.

As animals grow up, they must learn to find food. Young predators begin hunting lessons by watching the experts—mom and dad. The plant-eaters don't learn how to kill. They do learn where to find food at different times of the year.

A few kinds of animals almost always hunt with the help of animal friends. White pelicans, African wild dogs and wolves hunt with others of the same **species,** or kind.

Baby swans, called cygnets, learn what to eat and how to find it by watching mom

USING THE SENSES

Good hunting depends upon good senses. Predators such as big cats, wolves, foxes and coyotes have good eyes, sharp ears and a good sense of smell.

Eagles have little or no sense of smell, but they have amazing eyesight. Bears don't have great eyes, but their noses are terrific! Like raccoons, bears have a wonderful **tactile** sense, or sense of touch.

Amazingly quick for its size, an Alaska brown bear chases—and catches— a slippery sockeye salmon

CATCHING THE MEAL

Many animals have special and unusual ways to catch a meal. Long beaks and long noses can reach into flowers, mud or—in the anteater's case—termite mounds! The giraffe's long neck helps it reach leaves in tall trees.

Pelicans trap fish in their huge, pouched bills. Spoonbills slurp water and use feelers in their bills to find tiny animals. Flamingoes and some whales strain food from water.

Cheetahs dash at almost 70 miles per hour to catch prey. Many meat-eating birds dive at even faster speeds to attack prey.

A pirate in feathers, a gull perches on a pelican to steal any fish that may slip from the pelican's pouch

HUNTING WITH TOOLS AND LURES

A few animals have really unusual ways of getting food. Chimpanzees may poke a blade of grass into a termite's dirt mound. As termites rush to attack the stem of grass, the chimp pulls the stem back and picks off the insects.

The alligator snapping turtle has a built-in fish lure. The turtle opens its mouth and wiggles its pink, wormlike tongue. When a fish swims to the "worm"—chomp!

A sea otter cracks a clam against the rock it holds on its stomach

PREPARING THE MEAL

A lion eats its prey immediately. A bighorn sheep in a mountain meadow **grazes** anytime it's hungry. However, sometimes animals "prepare" their food.

The Egyptian vulture, for example, cracks ostrich eggs with rocks! Gulls break clams by dropping them onto rocks or roads. Sea otters use stones to break open shellfish.

Many animals, including wolves and storks, prepare meals for their youngsters by throwing up food they have already eaten.

Glossary

carnivore (KAR nuh vore) — a meat-eater

graze (GRAZE) — to feed on low-lying plants, especially grasses

herbivore (ERB uh vore) — a plant-eater

omnivore (AHM nuh vore) — an animal that eats both plants and animals

prey (PRAY) — an animal that is hunted by another animal for food

species (SPEE sheez) — within a group of closely-related animals, one certain kind, such as a *grizzly* bear

tactile (TAHK til) — relating to the sense of touch

INDEX